THE STORY BEHIND

SOAP

Christin Ditchfield

Heinemann Library
Chicago, Illinois

www.heinemannraintree.com
Visit our website to find out more information about Heinemann-Raintree books.

To order:
☎ Phone 888-454-2279
🖳 Visit www.heinemannraintree.com to browse our catalog and order online.

Edited by Megan Cotugno and Diyan Leake
Designed by Philippa Jenkins
Original illustrations © Capstone Global
    Library Ltd (2012)
Illustrated by Philippa Jenkins
Picture research by Hannah Taylor and Mica Brancic
Originated by Capstone Global Library Ltd
Printed and bound in China by CTPS

15 14 13 12 11
10 9 8 7 6 5 4 3 2 1

Library of Congress Cataloging-in-Publication Data
Ditchfield, Christin.
   The story behind soap / Christin Ditchfield.
     p. cm.—(True stories)
   Includes bibliographical references and index.
   ISBN 978-1-4329-5435-2 (hardcover)
   1. Soap—Juvenile literature. 2. Soap—History—Juvenile literature. 3. Soap trade—Juvenile literature. I. Title.
   TP991.D58 2012
   668'.12—dc22   2010044353

Acknowledgments
We would like to thank the following for permission to reproduce photographs: Alamy pp. **13** (© World History Archive), **14** (© Hemis), **15** (© North Wind Picture Archives), **20** (© Craig Lovell/Eagle Visions Photography), **25** (© Bon Appetit), **26** (© Osb70); Corbis p. **21** (Jacqui Hurst); Getty Images pp. **10** (AFP Photo/Leon Neal), **12** (The Bridgeman Art Library), **16** (Science & Society Picture Library), **17 bottom** (Boyer/Roger Viollet), **18** (Hulton Archive), **19** (Apic), **22** (Dorling Kindersley/Gary Ombler), **23** (Gary Ombler); iStockphoto p. **iii** (© Mihai Simonia); Shutterstock pp. **4** (7505811966), **5** (© Kasia), **6** (© Alexander Raths), **8** (© Muriel Lasure), **9** (© Elena Rostunova), **11** (© serg_dibrova), **17 top** (© Norberto Mario Lauria), **24** (© Ekaterina Znosok), **27** (© Hallgerd).

Cover photograph of four bars of hand-made, natural scented soap piled up on a white towel reproduced with permission of Alamy (© funkyfood London—Paul Williams).

We would like to thank Ann Fullick for her invaluable help in the preparation of this book.

Every effort has been made to contact copyright holders of material reproduced in this book. Any omissions will be rectified in subsequent printings if notice is given to the publisher.

Contents

Some words are shown in bold, **like this**.
You can find out what they mean by
looking in the glossary on page 30.

What Is Soap?

 Soap comes in different shapes, sizes, and colors.

What's in a name?

The word *soap* comes from the word *saipo*, which is what the ancient **Celts** called their version of this cleaning substance. The Celts lived in Britannia, or what is now the United Kingdom, around 500 BCE.

For thousands of years, people all over the world have been making and using soap. This gooey, slippery substance is used together with water to wash and clean things. Soap **dissolves** in water. It also dissolves in dirt, grease, and grime. When you wash something, soap joins onto the greasy dirt. It makes it easy to wash the dirt away with water.

Many different forms

Today, soap comes in many different sizes and shapes. It can be solid or liquid, hard or soft. Some soap comes in bars or squares. Soap also comes in flakes, powders, sprays, or foams. Dyes give soap its color. **Fragrant** oils and spices give it different scents.

How soap works

Soap **molecules** have two parts. One end is "water-hating" and sticks to grease and dirt. The other end is "water-loving" and dissolves in water. The soap molecules form a shell around the dirt. The water-loving parts of the soap molecules stick out. They dissolve in the water and carry the greasy dirt away.

Soap is everywhere

Soap can be found in Asia, Africa, Europe, and North and South America. Nearly every **culture** on every continent has its own version of soap. Over time, each **civilization** created its own recipe. They mixed herbs and plants, tree bark and branches, animal fats, and ashes from their cooking fires to create this powerful cleaning substance.

Soap Science

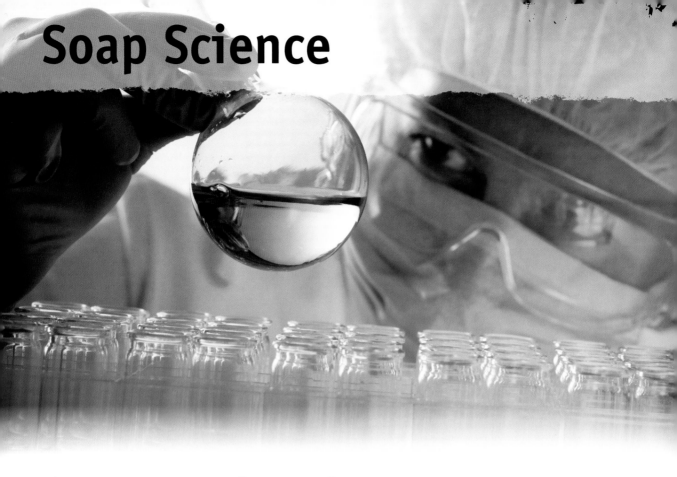

▲ **A research scientist studies the liquids that form soap.**

Soap is usually made from animal fats or vegetable oils. The fat is boiled with a **chemical** known as an alkali. Sodium hydroxide is the alkali often used in soap-making.

The science of soap

Some substances are made up of particles called **ions**. If a chemical has many **hydrogen** ions, scientists call it an **acid**. If a chemical has few or no hydrogen ions, it is called a **base**. When acids and bases are mixed together, they **react** to make new chemicals called salts. This is what happens when soap is made. An acid (from animal or vegetable fat) mixes with a base (also called an alkali) to create a salt that we know as soap.

Different kinds of soap have different qualities or characteristics. It depends what type of acid (fat) and what type of base (alkali) were used to make the soap.

On the job

A **chemist** is a person who studies substances to learn what they are made of and how they react (change when mixed) with other substances. Chemists study soap to understand how and why it works the way it does—and how it can be made even better.

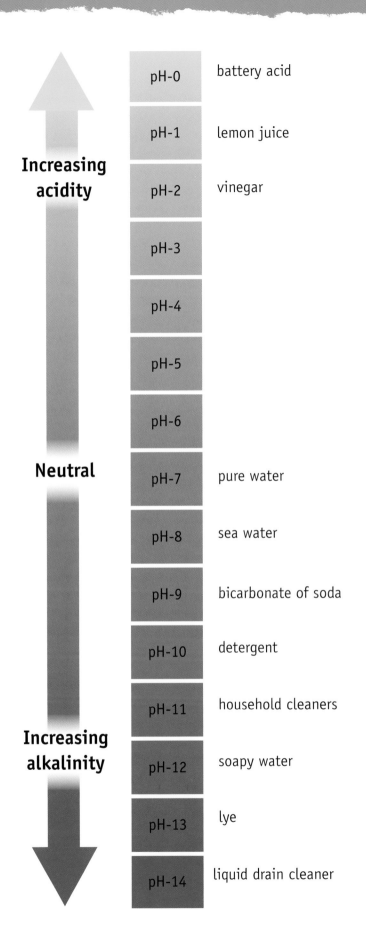

Increasing acidity

pH-0	battery acid
pH-1	lemon juice
pH-2	vinegar
pH-3	
pH-4	
pH-5	
pH-6	
pH-7	pure water
pH-8	sea water
pH-9	bicarbonate of soda
pH-10	detergent
pH-11	household cleaners
pH-12	soapy water
pH-13	lye
pH-14	liquid drain cleaner

Neutral

Increasing alkalinity

Then and now

For thousands of years, people made soap using a base called potassium hydroxide, or **lye**. Back then, lye was made from the ashes of burned plant and wood fibers. Today, nearly all soaps are made using lye that comes from laboratories. The chemical name for this lye is sodium hydroxide.

◀ Scientists use something called a pH scale to determine whether a liquid is an acid or a base.

What Soap Can Do

Soap makes our world cleaner and safer. It helps to kill the **germs** that can make us sick.

Cleaning with soap

We use soap to wash our clothes, our dishes, our toys, and our cars. We also use soap to wash ourselves. Soap cleans away all the dirt, sweat, and grime. It makes us smell better, too!

Special soap

Many soaps have added ingredients that give them different properties. These ingredients include oils such as olive oil, coconut oil, palm oil, cocoa butter, and shea butter. They also include things such as aloe (a plant that is used to treat minor wounds and burns), honey, almonds, or rolled oats. These special soaps can make your skin feel very soft and smooth. They often smell very nice, too.

▼ Some soaps are made with special ingredients.

Tough soaps for tough jobs

Industrial soaps are extremely strong soaps used to clean up heavy-duty **chemicals** and waste. They are much too harsh—too tough—to use on fabrics or human skin.

When to wash

You should always wash your hands:

- before cooking
- before eating
- after using the bathroom
- after cleaning up around the house
- after touching animals, including your pets
- before and after visiting any sick friends or relatives
- after blowing your nose, coughing, or sneezing
- after being outside.

▲ Fifty people are being fit into this giant bubble. This was a world record set in 2007.

Fun with soap

Soap can be a lot of fun—especially when it is used to make bubbles! Water bubbles are made of air, trapped by a thin film or skin of water. They pop almost immediately. When you add soap to water, the bubbles last longer. The soap makes the film around the bubble stronger and stretchier.

The world's largest bubble

On July 30, 2009, an entertainer named Samsam Bubbleman created the world's largest and longest free-floating bubble. It was 500 feet (152 meters) long!

Activity

Make your own soap bubbles

You will need:

2/3 cup of liquid dishwashing detergent
1 gallon (4 liters) of water
2–3 tablespoons of glycerin

or

1 cup of liquid dishwashing detergent
8 cups of water
4 teaspoons of sugar

1. Mix the ingredients together in a large bowl or bucket. Stir gently! Don't let it get foamy, because then it won't work.

2. If you can, let it sit overnight. The bubbles will be even better!

3. Find some bubble wands, pipe cleaners bent into different shapes, straws, six-pack rings, funnels, or string (tied in loops onto straws or pipe cleaners).

4. Dip them into the bubble solution and then blow gently.

Hint: Glycerin makes the bubbles really strong—but if you can't find it, use the other recipe. Sugar works in a similar way.

A Short History of Soap

▲ **The ancient Egyptians used soap to treat their skin.**

No one knows for certain who first discovered soap or how it was invented. Historical records do show that as early as 2800 BCE, the Babylonians (people who lived in the place we now call Iraq) were making soap. Writings on an old **papyrus** reveal that the ancient Egyptians used soap to treat skin diseases in 1500 BCE. The Phoenicians (who lived along the coast of what we now call Lebanon, Syria, and Israel) used soap to wash wool and cotton to prepare it for weaving in 600 BCE.

Keeping clean

At first, people used soap only to clean their clothes. They cleaned themselves by scraping the dirt off their skin. Some used rocks, shells, or tools that looked like sticks. Others used a mixture of oil and sand or gritty clay that they rubbed against their skin.

The Greeks, the Romans, and the Celts

A doctor in ancient Greece named Galen recommended that people wash with soap to prevent the spread of disease. Another wrote about the use of soap as shampoo. Both the Romans and the ancient **Celts** were known to use *sapo* or *saipo*. Each claimed that the other had learned to use it by copying them.

> **Soap story**
>
> According to **legend**, the women of Rome used to wash their clothes in the Tiber River at the bottom of Sapo Hill. They noticed that the clothes seemed cleaner than those washed in any other river. They discovered that the ashes and grease from animal sacrifices made in the temple at the top of the hill had mixed with the rain and run down into the river. There it became the soap that cleaned their clothes.

▼ The Greek doctor Galen gave lectures on medicine and health.

13

▲ Soap made in factories was cut into large blocks and stamped with the name of the manufacturer or the region where it was made.

Medieval times

In the **Middle Ages**, soap-makers (also called soap-boilers) produced large quantities of soap in factories all across Europe, Africa, Asia, and the Middle East. They organized themselves into guilds, or business groups. These soap-makers added perfumed oils, herbs, and spices to their soaps to make them smell better. They also used vegetable dyes to give their soaps color. Liquid soap was poured into square frames and left to harden in large blocks. Merchants then cut off the amount of soap a customer wanted to purchase and charged them by the weight.

Colonial times

Only the wealthy could afford soaps made in factories. Ordinary families still made their own soap at home or purchased it from the village soap-maker. This was certainly true for those living in **colonial America** in the 1600s.

They took ashes from their wood fires and boiled them with water and animal fats, until the water boiled away. They added more ashes and water and boiled the mixture again. They repeated this process until the fats broke down and **reacted** with the ashes to form soap. It was a long, difficult, and messy job.

▼ **This image shows people preparing fat and boiling it to make soap.**

▲ Soap manufacturers began producing soap in small, square-shaped bars for the first time in the late 1700s and early 1800s.

The Industrial Age

In the 1800s, scientists begin studying the **chemical** process of soap-making. They asked questions such as: Why do these ingredients **react** the way they do? What is in the fats and the ashes that cause this reaction? The answers to these questions changed the process of soap-making for ever. Once soap-makers understood how and why the process worked, they could make greater improvements to it.

Important discoveries

In 1791, a French inventor named Nicolas LeBlanc discovered a way to create a new kind of **lye**. It was called caustic soda ash and was made from sea salt instead of the ashes of wood fires. This was important because Europe was running out of trees to burn! A Belgian **chemist** named Ernest Solvay improved LeBlanc's method, creating soda ash from brine (salt water) and limestone (a kind of mineral rock).

In 1823, a French chemist named Michel Eugène Chevreul discovered that the soap-making process released a substance from the animal and vegetable fats. He called this substance glycerin. Glycerin could be left in the soap or it could be removed. If it was removed, glycerin could become an important ingredient in all kinds of other products—sweets, cosmetics, printing ink, and even dynamite (nitroglycerin)!

▼ Glycerin can be used to make a clear, all-natural moisturizing soap.

Michel Eugène Chevreul (1786–1889)

Michel Eugène Chevreul made many important scientific discoveries, including the properties of glycerin. He worked at the Museum of Natural History in Paris, France, and it was there that he began his studies into animal fats in 1811. Chevreul later invented a new kind of candle that burned brighter than other types. He lived until the age of 102!

To lecture or preach to others on a subject you are passionate about is called "getting on your soap box."

Modern times

In the 1900s, soap became very easy to make and buy. During World Wars I and II, chemists began experimenting with ways to make imitation soaps. (They needed the chemicals normally used to make soap to be available for other uses, such as making explosives.) These substances look like soap and smell like soap and work like soap, but they are made of completely different chemical ingredients. These imitation soaps are called **synthetic detergents.**

In the last 50 to 60 years, hundreds of new kinds of soap and detergent have been invented, including deodorant soap, liquid hand soap, **antibacterial** soap, and **hand sanitizers**.

Soap operas

In the 1930s, soap-makers advertised their cleaning products when daytime radio and televison drama programs were broadcast. They wanted to reach housewives and homemakers who might use the soap. The programs themselves came to be known as soap operas.

▼ Soap is a beauty product as well as a cleaning substance. This advertisement is from 1958.

How Soap Is Made

▲ Hot liquid soap is poured into molds, where it will cool and harden.

Today there are two main methods of making soap: the hot process and the cold process.

Hot process

In the hot process method, soap-makers boil the fats or oils and the **lye** together in a very large kettle, at a very high temperature. A **chemical** reaction called **saponification** takes place. The individual ingredients join together to become a new substance—soap—along with glycerin. Soap-makers remove the glycerin and any excess water from the surface.

The hot liquid soap is poured into molds. As the soap cools, it hardens. Salt water can be used to make the soap harden faster and better.

Cold process

In the cold process method, soap-makers **dissolve** the lye in water without heating it. They melt the fat until it becomes a liquid, but they don't boil it. They stir the lye and the fat together until the mixture has the same texture as poster paint. The liquid is poured into molds and left to sit for up to 48 hours. While the liquid sits in the molds, the process of saponification takes place. When the soap is removed from the molds, it is left to dry for two to six weeks.

◀ There are many small factories where soap is still made by hand, rather than machine.

▲ Most soaps are made in large factories using specially designed machines like this one.

The purification process

The soap-making process is not complete until the soap has been **purified**. Certain **chemicals**, such as sodium chloride, sodium hydroxide, and glycerol, need to be removed. The purification process takes place once the liquid soap has dried. Soap-makers boil the new soap in water over and over again, until all the **impurities** float to the surface. They drain the water and then add salt to the soap to help it harden.

From there, the soap can go back into a drying rack where it will dry over time. The soap can also be put into a vacuum chamber (container from which most of the air has been removed) or under a spray dryer until it is quickly and thoroughly dried out. Once the purification process is finished, the soap can be molded into bars or shaved into flakes and powders.

Deep clean

Soap-makers often add perfumes, colors, moisturizers, and medicines to their soaps. Some manufacturers also add other ingredients such as sand or pumice (volcanic rock). These gritty substances work like sandpaper to exfoliate (scrape away) dry, dead skin cells and leave fresh, smooth skin underneath.

Stir it up!

When soap is stirred vigorously in water or rubbed back and forth against your skin, it creates a bubbly foam called lather. The amount of lather depends on what types of oils were used to make the soap and whether the soap also includes salt, milk, or sugar. These ingredients make the bubbles bigger and the foam thicker.

▼ **This machine produces soap in small bars for ordinary, everyday use.**

Soap as Art

▲ These glycerin soaps were made by hand—and shaped into bars—in a specialty soap-making shop.

Today people all over the world make soap for the same reasons people like to cook or knit. It's a great **hobby**! Soap-makers enjoy creating unusual colors, shapes, textures, and fragrances. The more skilled the artist, the more elaborate (detailed) the design. Some artists sell their homemade creations at craft fairs or in gift shops.

Activity

Make your own soap

You will need:

3 cups of soap flakes
1½ cups of water
liquid food coloring
vegetable oil
a large bowl

1. Pour the water and the soap flakes into the bowl.
2. Add a few drops of food coloring.
3. Mix these ingredients together with your hands, until it looks and feels like playdough.
4. Rinse the mixture off your hands and put a few drops of vegetable oil on them to make them slippery.
5. Form the dough into whatever shapes you like. You can also use a cookie cutter or a Jell-O mold.
6. Let the soap dry overnight. You can use it the next day!

*Hint: If you can't find soap flakes, you can make your own by grating a bar of soap. To make soap from scratch, you need to work with **chemicals** and will require adult supervision.*

The Future of Soap

▲ Although many people now use detergents and other kinds of cleansers, natural handmade soaps are becoming very popular once again.

Today most of the products that we call soap are actually **detergents**. Detergents are cleaning solutions that act like soap, but are made from different **chemical** ingredients. In many ways, these new detergents are much better soaps. They work better. They clean better. They are cheaper and easier to make, and easier to use. They don't leave a sticky **residue**—soap scum—on glass or other surfaces.

Some of the chemicals in these detergents may not be safe for the environment. Some can be irritating to sensitive skin. Others can cause **allergies**. For these reasons, old-fashioned soap is becoming popular again. Many people are choosing to use it as part of a healthier, more organic (natural) lifestyle.

Soap to the rescue! ✔

Since the 1970s, liquid dishwashing soap has been used to help clean up oil spills that threaten the environment. The soap breaks down the oil so that it can be more easily removed from oceans and beaches. Animal rescue teams use soap to help gently remove oil from the feathers, fur, and skin of oil-soaked birds and other creatures.

Official definition ✔

The definition of true soap is a product made from the reaction of sodium hydroxide, fats, and oils. A cleaning product made from any other ingredients is a detergent or cleanser.

Timeline

(These dates are often approximations.)

2800 BCE
The Babylonians create soap by boiling animal fat and mixing it with ashes.

3000 BCE →

100 CE
The Greek physician Galen recommends that people use soap to wash themselves (and not just their clothes), to prevent the spread of disease.

79 CE
The volcanic eruption of Mount Vesuvius in Italy destroys the city of Pompeii, including its soap factory.

200 CE ← 100 CE ← 0

385 CE
The Roman physician Priscianus describes the practice of using soap as shampoo.

300 CE → 400 CE

1100 ← 1000 ← 900 CE

1000–1400
During the Middle Ages, large quantities of soap are produced in factories and towns all across Europe—in Bristol, Castile, Marseilles, Savona, and Venice.

1200 → 1300

1811
French **chemist** Michel Eugène Chevreul obtains the first **patent** on glycerin.

1700s–1800s
Scientists begin studying the chemistry of soap, experimenting with new ingredients. The first bar soaps are made.

1900 ← 1800 ← 1700

1914
During World War I, the Germans create some of the first **synthetic** soaps, so that they can use animal and vegetable fats for other purposes.

1950s
During World War II, the most common ingredient used in the production of synthetic soaps and **detergents** is a **chemical** called alkylbenzene.

28 ⋀⋀⋀ This symbol shows where there is a change of scale in the timeline, or where a long period of time with no noted events has been left out.

1500 BCE

The ancient Egyptians use soap made of both animal fats and vegetable fats to treat skin diseases.

2000 BCE ——————————————→ 1000 BCE

300 BCE–300 CE

Both the Romans and **Celts** use *saipo*, or soap. Both **cultures** claim to have invented it and shared it with the other.

600 BCE

The Phoenicians use soap to clean wool and cotton, to prepare it for weaving into cloth.

←———————————————

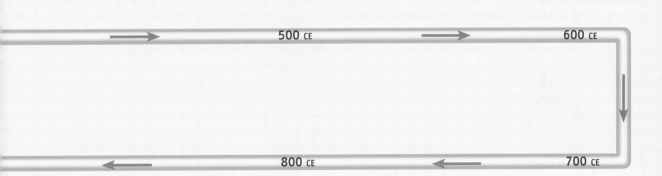

————————————→ 500 CE ————————————→ 600 CE

←———————————— 800 CE ←———————————— 700 CE

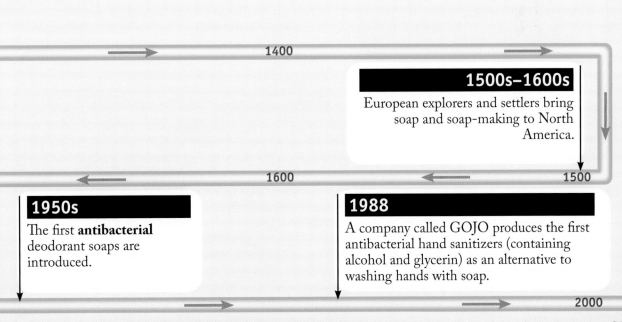

————————————→ 1400 ————————————→

1500s–1600s

European explorers and settlers bring soap and soap-making to North America.

←———————————— 1600 ←———————————— 1500

1950s

The first **antibacterial** deodorant soaps are introduced.

1988

A company called GOJO produces the first antibacterial hand sanitizers (containing alcohol and glycerin) as an alternative to washing hands with soap.

————————————————————→ 2000

Glossary

acid liquid with many hydrogen ions; when mixed with a base, it creates sodium (salt)

allergy unusual reaction (such as itching, sneezing, or rashes) to certain substances

antibacterial having ingredients that fight and kill bacteria or other kinds of germs and infections

base liquid with few or no hydrogen ions; when mixed with an acid, it creates sodium (salt)

Celt member of a group of people who lived in Britain around 500 BCE

chemical substance that can be made into other substances by changing its atoms or molecules. Chemicals may be found in nature or made in laboratories.

chemist person who studies substances to learn what they are made of and how they react with other substances

civilization culture of a particular time or place

colonial America period when America was a British colony with English settlers, before the United States gained independence in the 1700s

culture set of shared attitudes, values, goals, and practices that a group of people has

detergent chemical substance that works like soap to clean things

dissolve break down or disappear in a liquid (such as water), becoming part of the liquid

fragrant having a sweet or pleasant smell

germ tiny living thing that can cause disease

hand sanitizer antibacterial liquid used to clean hands without washing with soap and water

hobby something a person enjoys doing in their spare time

hydrogen simplest and lightest of all chemical elements

impurity something that does not belong in a particular substance

ion atom or group of atoms that has an electric charge

legend ancient story that is told over many generations

lye liquid base made from the ashes of burned plant and wood fibers

Middle Ages period of European history from 500 to 1500 CE

molecule tiny particle composed of one or more atoms

papyrus thin paper made from the stems of the papyrus plant

patent rights to an invention for a limited period of time. The patent allows the owner to make money from an invention without competition from other companies.

purify make something pure and clean

react change due to a chemical process

residue something left behind, when the main part has been taken away

saponification process by which fats, oils, and lye become soap

synthetic made from chemicals, not found in nature

Find Out More

Books

Browning, Marie. *Totally Cool Soapmaking for Kids*. New York: Sterling, 2005.

Rhatigan, Joe. *Soapmaking: 50 Fun and Fabulous Soaps to Melt and Pour*. Asheville, NC: Lark Books, 2005.

Stein, David. *How to Make Monstrous, Huge, Unbelievably Big Bubbles*. Palo Alto, CA: Klutz, 2005.

Websites

http://www.artistshelpingchildren.org/barsofsoapcraftsideasdecorationskids. html
This site features basic instructions on soap sculpting and more than fifty different craft projects you can make using soap.

http://www.earthskids.com/basic_handwashing_info-kids2.htm
This website has a special page full of soap-themed science experiments, links, and resources for kids.

www.teachsoap.com
This site has all kinds of information on soap-making. The kids' page includes recipes for "fizzy bath bombs" and other fun soap projects.

Index